THE LAST SPECIES
ENDLINGS

Northern White Rhino

by Joyce Markovics

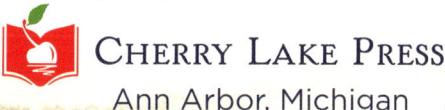

Cherry Lake Press
Ann Arbor, Michigan

Cherry Lake Press

Published in the United States of America by Cherry Lake Publishing Group

Ann Arbor, Michigan

www.cherrylakepublishing.com

Reading Adviser: Beth Walker Gambro, MS Ed., Reading Consultant, Yorkville, IL

Content Adviser: David R. Sischo, PhD, Wildlife Biologist

Book Designer: Ed Morgan

Photo Credits: Make it Kenya Photo/Stuart Price/flickr, cover and title page; © Steve Tum/Shutterstock, 5; Ray in Manila/flickr, 6; © J.A. Dunbar/Shutterstock, 7; © JONATHAN PLEDGER/Shutterstock, 8; © neelsky/Shutterstock, 9; © Kauka Jarvi/Shutterstock, 10; © Francois van Heerden/Shutterstock, 11; © JONATHAN PLEDGER/Shutterstock, 12; freepik.com, 13; © Holly Auchincloss/Shutterstock, 14; © Trebor Eckscher/Shutterstock, 15; © REUTERS/Alamy Stock Photo, 16–17; © Design_Cells/Shutterstock, 18; freepik.com, 19; Make it Kenya Photo/Stuart Price/flickr, 20–21; © Mehdi Kasumov/Shutterstock, 22 top; Wikimedia Commons, 22 middle and bottom.

Copyright © 2023 by Cherry Lake Publishing Group

All rights reserved. No part of this book may be reproduced or utilized in any form or by any means without written permission from the publisher.

Cherry Lake Press is an imprint of Cherry Lake Publishing Group.

Library of Congress Cataloging-in-Publication Data

Names: Markovics, Joyce L., author.
Title: Northern white rhino / by Joyce Markovics.
Description: Ann Arbor, Michigan : Cherry Lake Publishing, [2023] | Series: Endlings. The last species | Includes bibliographical references and index. | Audience: Grades 4-6
Identifiers: LCCN 2022004103 (print) | LCCN 2022004104 (ebook) | ISBN 9781668909669 (hardcover) | ISBN 9781668911266 (paperback) | ISBN 9781668912850 (ebook) | ISBN 9781668914441 (pdf)
Subjects: LCSH: White rhinoceros—Juvenile literature. | White rhinoceros—Conservation—Juvenile literature. | Endangered species—Juvenile literature.
Classification: LCC QL737.U63 M335 2023 (print) | LCC QL737.U63 (ebook) | DDC 599.66/8–dc23/eng/20220315
LC record available at https://lccn.loc.gov/2022004103
LC ebook record available at https://lccn.loc.gov/2022004104

Printed in the United States of America by Corporate Graphics

THE LAST SPECIES CONTENTS

A Great Loss 4
Rhino Facts 8
Racing Against Time 14
Science at Work 18

    Animals Under Threat 22
    Glossary 23
    Find Out More 24
    Index 24
    About the Author 24

A Great Loss

On March 19, 2018, the last male northern white rhinoceros died in Kenya, Africa. His name was Sudan, and he was 45 years old. Sudan's huge, old body had failed, leaving him unable to walk. He fell to the ground like a giant gray boulder.

Before he died, Sudan's keepers stroked his huge, horned head. They spoke softly to him in Swahili, their **native** language. Despite his size, Sudan was gentle and sweet. His keepers knew they were not only losing the last male northern rhino—but also a friend.

Sudan was the last male northern white rhino. The last known animal of its kind is called an endling.

Sudan was not the last northern white rhino. He's survived by his daughter, Najin, and his granddaughter, Fatu. However, without a male, the female rhinos can't have young naturally.

Sudan's grave in Kenya

Scientists have a term for when there are too few animals for a **species** to survive. It's called functional **extinction**. However, scientists have not lost all hope. They're trying to save northern white rhinos from disappearing forever. Whether it will work has yet to be seen.

Najin and Fatu, the last two northern white rhinos

Sudan was the last northern white rhino that lived in the wild. Both Najin and Fatu were born in a zoo.

Rhino Facts

Rhinos are ancient giants. They first walked the Earth 55 million years ago. Today, there are five different kinds of rhinos. Three rhino species are native to Asia. The two others—black rhinos and white rhinos—live in Africa. White rhinos include northern and southern varieties.

A southern white rhino in South Africa

This Indian rhino only has one horn and lives in India and Nepal.

All rhinos have a few things in common. They're large, heavyset **mammals** with strong legs. They have thick, folded gray skin. Rhinos have poor eyesight and either one or two horns on their enormous heads. The horns are used for **defense** or looking for food.

African white rhinos and black rhinos are actually the same color—brownish gray!

Northern white rhinos are the largest rhino species. They stand around 6 feet (1.8 meters) tall and weigh 5,000 pounds (2,268 kilograms). That's about as much as a Jeep. These creatures have wide mouths for grazing on grasses. In fact, they're eating machines.

A white rhino wallowing in mud

Rhinos spend most days and nights munching on grass. The rest of their time is spent resting or **wallowing** in mud. The cool mud feels good on their thick yet sensitive skin. It also protects rhinos from sunburn and bug bites.

Northern rhinos eat more than 120 pounds (54 kilograms) of grass per day. And they produce cantaloupe-size balls of poop!

Despite their massive size, northern white rhinos are fast. These huge animals can charge at speeds of 40 miles (64 kilometers) per hour. It's no wonder that a group of rhinos is called a "crash." Because of their poor eyesight, rhinos startle easily and may charge at anything—even a tree!

This young rhino is charging at full speed.

A mother rhino and her calf

Northern white rhinos often live alone or in small groups. Every two to five years, male and female rhinos come together to **mate**. It can take 16 months before a female rhino gives birth. The baby, or calf, will stay with its mother for up to 4 years.

> The northern white rhino is the second-largest land mammal in the world. Elephants are the biggest.

Racing Against Time

Rhinos have one deadly enemy—humans. For hundreds of years, people have hunted rhinos for their horns. In some places, powdered rhino horn is used to treat illnesses. Yet rhino horns are made from the same material as human fingernails. Like fingernails, the horns have no healing ability.

This rhino was killed for its horn, which has been cut off.

This hasn't stopped people from hunting rhinos, especially northern white rhinos, to extinction. Even after governments worked to protect rhinos, **poachers** continue to kill them for their horns.

The skeleton of a poached rhino in Africa

People have also taken over and destroyed the habitats where rhinos live. This, too, has caused their decline.

Poachers are still a threat to the last two northern white rhinos in the world. As a result, armed guards protect Najin and Fatu around the clock. But the rhinos don't seem to notice.

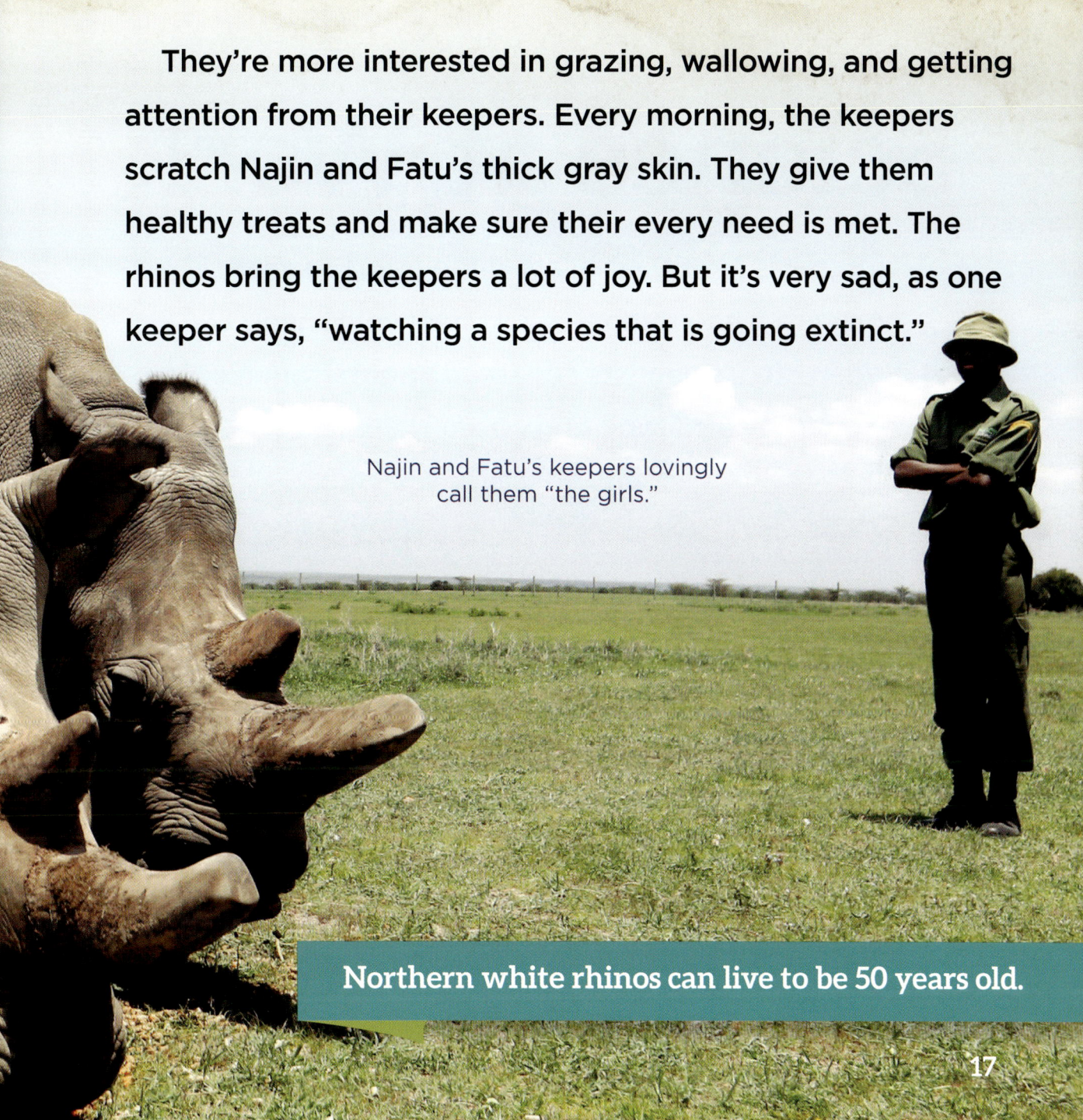

They're more interested in grazing, wallowing, and getting attention from their keepers. Every morning, the keepers scratch Najin and Fatu's thick gray skin. They give them healthy treats and make sure their every need is met. The rhinos bring the keepers a lot of joy. But it's very sad, as one keeper says, "watching a species that is going extinct."

Najin and Fatu's keepers lovingly call them "the girls."

Northern white rhinos can live to be 50 years old.

Science at Work

Experts are trying a last-ditch effort to save northern white rhinos—using science! When Sudan died, scientists collected **sperm** from his body. They also took **eggs** from Najin and Fatu.

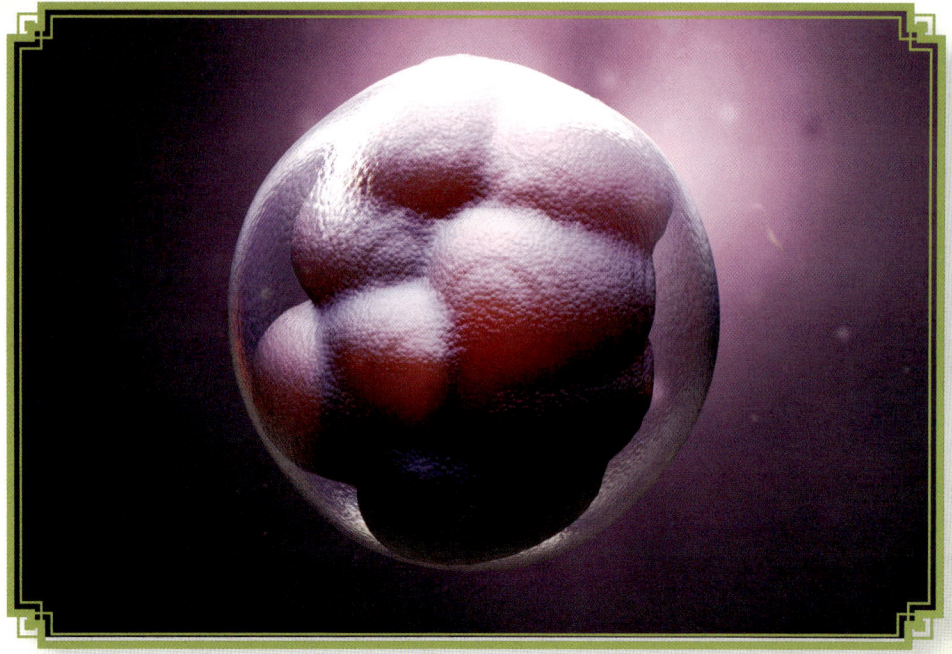

This is what an embryo looks like under a microscope.

A scientist stores embryos in an extra-cold substance called liquid nitrogen.

They joined the sperm and eggs together in a lab and made embryos. An embryo is a baby at a very early stage of development. Then they froze the embryos. Freezing can **preserve** an embryo for many years.

The embryos are stored in a special freezer at the San Diego Zoo.

The scientists are planning to put the embryos into female southern white rhinos. Najin and Fatu have health problems, so they can't become mothers. The hope is that the embryos will grow in the southern white rhinos.

Then new northern white rhino calves will be born. And they might help save the species. Paul Baribault, who leads the program, says, "Our **ultimate** goal is not to create a single northern white rhino but reintroduce an entire herd." However, the first step is one northern white rhino calf.

Animals Under Threat

Many more large mammals are at risk of dying out. Here are three on the brink of extinction:

Javan Rhinoceros
This one-horned rhino is one of the rarest large mammals on Earth. Only around 75 exist.

Mountain Gorilla
These powerful great apes live in central Africa. There are less than 1,000 mountain gorillas remaining.

Tiger
Tigers are native to the forests of Asia. There are fewer than 4,000 of these big cats left in the world.

Glossary

defense (di-FENSS) the act of protecting oneself

eggs (EGGZ) female reproductive cells

extinction (ek-STINGK-shuhn) when a certain living thing dies out completely

habitats (HAB-uh-tats) places in the wild where animals normally live

mammals (MAM-uhlz) warm-blooded animals that have hair or fur and nurse their young

mate (MATE) to come together to have young

native (NAY-tiv) belonging to a particular place

poachers (POH-chuhrz) people who hunt animals illegally

preserve (pri-ZURV) to keep and protect something

species (SPEE-sheez) certain types of animals or plants

sperm (SPERM) a male reproductive cell

ultimate (UHL-tuh-muht) final

wallowing (WAH-loh-ing) rolling around in something such as mud

Find Out More

Books

Hoare, Ben, and Tom Jackson. *Endangered Animals*. New York, NY: DK Children, 2010.

Riera, Lucas. *Extinct: An Illustrated Exploration of Animals That Have Disappeared*. New York, NY: Phaidon Press, 2019.

Whitfield, John. *Lost Animals*. New York, NY: Welbeck Publishing, 2020.

Websites

National Geographic: The Photo Ark
https://www.nationalgeographic.org/projects/photo-ark/

San Diego Zoo Wildlife Alliance
https://science.sandiegozoo.org/species/white-rhino

Six Extinctions: An Overview of the Ends of Species
https://www.amnh.org/shelf-life/six-extinctions

Index

Africa, 4, 8–9, 15, 22
Asia, 8, 22
embryo, 18–20
endangered animals, 22
extinction, 7, 15, 22
Indian rhinoceros, 9
mammals, 9, 13, 22
northern white rhinoceros,
  calves, 13, 20
  charging, 12
  diet, 10–11

Fatu, 6–7, 16–20
habitat, 8–9, 15
horns, 4, 9, 14–15, 22
keepers, 5, 16–17,
mating, 13
Najin, 6–7, 16–20
Sudan, 4–7, 18
wallowing, 11, 16
poachers, 15–16
southern white rhinoceros, 8, 20
zoo, 7, 19

About the Author

Joyce Markovics has written hundreds of books for kids. She hopes this book inspires young readers to learn more about endangered animals and take action to prevent their extinction. She dedicates this book to Ava, a fellow animal lover with an inquisitive mind.